Communication skills for engineers and scientists

Third edition

Edited by John Venables

Published by
Institution of Chemical Engineers (IChemE),
Davis Building,
165-189 Railway Terrace,
Rugby, Warwickshire CV21 3HQ, UK
IChemE is a Registered Charity

First edition 1990
Second edition 1994
Third edition 2002

ISBN 0 85295 455 7

Cartoons by Bob Cheshire
Email: bob.cheshire@ntlworld.com

Printed in the UK by Page Bros, Norwich

Introduction

Good communication is vital in engineering and science. You need to be able to express concepts and ideas to colleagues. Information must be given to customers, the people you work with, and those who work for you — and they are not all engineers or scientists.

You may need to present a good case to senior managers. If you are appointed to lead a project team, you must be sure that everyone knows what is going on so that they can work efficiently. You might be required to give a media interview, or address a public meeting.

All of these challenges may sound daunting, but it is important to realize that good communicators are made, not born. All it takes is a little knowledge, which can be learned, and practice, which will come with experience.

Whatever your objective, this book will help you to get the message across, effectively and efficiently. It will introduce you to the communication tools at your disposal, explain how body language works and explore why cultural differences can cause breakdowns in communication, and lots more.

One final thought before you turn the page. Remember that communication is a two-way process. Knowing how to listen can be just as important as being able to host a multi-point video-conference or address the chairman of the board.

Good luck!

John Venables
2002

Contents

1. The basics

We live in a communication age. Telephones, faxes and mobiles reach distant places and people in seconds. Email and the web allow effortless transmission of words, pictures and sounds around the globe.

Yet communication tools will not automatically make you a good communicator, any more than owning a set of kitchen pans will turn you into a master chef.

Communication is not just about transferring ideas and information from one place to another. It is about putting information and concepts across in such a way that other people understand what you mean.

You need to select the right communication tool. But you must also craft your message in such a way that people grasp what you are trying to say quickly and easily, with the least possible chance of misunderstanding.

The tool kit

What do you want to do? Choose the most appropriate communication method for the job. For example:

Purpose	Communication method
Simple question / request	Telephone
Send complex instructions	Email with attachments / fax
Formal communication	Letter
Write to a few people	Email
Keep a team or section informed	Briefing / video-conference
Speak to the general public	Media interview
Make a proposal to the board	Audio-visual presentation

Making words count

Whatever method you choose, fax, mobiles, emails and so on are just vehicles for the ideas and information you want to communicate. They won't help if no-one understands what you are trying to say. The key to good communication is ensuring you have a clear message, and that means using words efficiently and effectively.

Vocabulary

Some of us need no more than 600–800 words in daily life, while an average educated person may use 5000 words as a matter of course.

A demanding profession will require additional specialized terms, which must be used precisely in context.

We tend to use a wider range of words in writing than in speech. As we move from speaking to writing, our vocabularies should increase, partly because the day-to-day shorthand of speech is no longer appropriate, and partly because you must introduce a greater variety of expression to avoid boring your reader.

Be adventurous. If you come across a new word, look up its meaning and mentally file it away for future use. Over time your vocabulary will broaden and ripen, allowing you to express yourself in new and richer ways.

Two types of publication can help you to improve the way you use words. Dictionaries will give you the precise definition of a word or term. A thesaurus will offer you alternatives to words, so enabling you to avoid constant repetition.

Grammar and punctuation

A certain amount of informality is acceptable, but it's important to remember that the way you link words can be as important as the words you use.

Correct use of the basic rules of language will help ensure that everyone understands you. It will also give others the impression that your message has been carefully crafted and thought through with proper attention to detail.

Use alternative words to avoid potentially irritating repetition

If you think of composing your message as building a wall, then the shape of the final message depends not just on the words (the bricks) but also on how you put them together (the construction method), and that means grammar, punctuation and syntax.

Plain language

Different professions, trades and branches of science and engineering have developed specialist languages to communicate particular ideas and concepts. That's fine if everyone involved in the conversation understands the terms used, but do realize that others may find the jargon you use wholly impenetrable.

For instance, a biochemist might have difficulty following a technical discussion between two electronics engineers, and vice versa. And both sides would have difficulty making themselves understood to a member of the general public.

Communication is about understanding. If you are talking to your peers then by all means use your private language. But if you have to communicate your subject to someone from another field or a non-specialist, then use everyday English — the language people speak in the street, in shops and at social gatherings.

Presentation

Whether you are using the telephone, speaking to people face-to-face or giving a talk, your audience is going to judge you not only on what you say but also how you say it. Make the most of your voice:

- Speak clearly and don't mumble or swallow your words.
- To err is human — but too many 'errs' and 'ums' may irritate and distract.
- Don't speak too fast — people will find it difficult to keep up with you.
- Clarify what you want to say in your mind before you open your mouth — you will sound more professional and authoritative.

Effective communication

The words you use and the way you use them are vital. But planning and preparation are also important.

For instance, before initiating a message, ask yourself:

- Who needs to know and when?
- What do they need to know?
- How should they receive the information and how often?

If you are on the receiving end of a communication, ask yourself:

- What is the purpose of the information you have received?
- Why you?
- What are you expected to do next?
- How do you retain / retrieve the information when required?
- What else do you need to know / check before you can act?

Keeping a record

Always make a note of any communication — it is all too easy to forget important details.

Log any telephone conversations, noting the time, who you spoke to, and what was agreed. If necessary, ask for written confirmation — for example, 'Would you mind dropping me a line confirming that price/delivery date, please?'

Emails log details of the sender and the time and date sent automatically. Don't be tempted to archive old emails too soon — you may need them again at short notice.

Use your computer diary to keep notes or, if you deal with many enquiries, use a daybook, with columns for date, time, message and action. Some companies may supply you with proprietary software to do this. And don't spurn the old-fashioned paper diary, Filofax® or shorthand or reporter's spiral bound notebook — at least it won't crash just when you most need it!

Top tips – the basics

- ✔ Think through what you want to communicate
- ✔ Choose the right tool for the job
- ✔ Use language people will understand
- ✔ Keep a record of important communications
- ✔ Be concise — time costs!

2. The written word

Whether you are writing a letter, instructions, an article or a report to a client, first decide what you want to say. Ask yourself, what does the recipient need to know if s/he is to understand the message?

Once you have decided on your message, structure the communication so it is easily readable and has maximum impact.

When writing a letter:

- State the purpose of your letter clearly.
- Highlight particular points that are essential to your argument / sales pitch / request.
- Ask for any further information you need from the recipient.
- Set out your conclusion.
- Suggest further action and who is responsible for carrying it out.
- Add the appropriate signing off phrases — 'Regards' for someone you know, 'Yours sincerely' is rather more formal (Dear Dr, Mr or Ms) or 'Yours faithfully' is very formal (Dear Sir / Madam).

When writing a report or appraisal, make sure you include:

- Distribution list.
- Title.
- Summary.
- Background.
- Objectives.

- Techniques used.
- Results.
- Conclusions / recommendations.
- Further information, references, acknowledgements, etc.

Try to stick to accepted formats — a sloppy letter or report will not impress and can even confuse. Most organizations have an accepted house style. If you are not certain how to compose your communication, ask for advice. Word processing software packages such as Microsoft Word® offer templates to help you prepare reports, letters, theses and other publications, but these may need customizing to fit your organization's requirements.

General points

- Whatever you are writing, stick to the point. Waffle will obscure your central message and irritate the reader.
- Read back what you have written. Does it say what you mean? If you are unsure, get colleagues to read the document through and get them to tell you what they think you are saying.
- Cut out superfluous material.
- Check for repeated words and phrases; find alternatives in the thesaurus. But don't get too carried away, as this can confuse readers for whom English is a second language.
- Use your software's spell and grammar checker, but don't trust them too far. For example, both 'unclear' or 'nuclear' would be passed by your spell-check, but the computer won't know which word you intended to use.
- Make sure you have got your facts right.

- Try to ensure that what you write cannot be taken the wrong way.
- Dry wit doesn't always work on the printed page. Put yourself in the position of the reader. Re-read what you have written, and if you feel threatened or insulted, the letter is over the top.
- Never write in anger — you may regret it later! Give yourself time to cool off before committing pen or printer to paper.
- Remember that written material of all kinds is on the record and can be used as hard, even legal, evidence of an agreement or discussion. If what you write is confidential, say so.
- Keep copies, and create a good filing system so that you can always retrieve information when it is needed. Back up hard disk files.
- Remember that for people who do not know you personally, what you write is you!

Notes, memos and letters: do's and don'ts

- Check for errors, sense and spelling before you send documents out.
- If you need to go into details, put them on a separate sheet as an attachment.
- Make sure you send copies to the people who need to know (attach a list of 'Copies to').
- Don't handwrite letters or memos — they will look amateurish and sloppy if your handwriting is poor.

Technical papers, articles and reports: do's and don'ts

- Good presentation is very important. It is in your interest to work hard on your report. If you are writing a report or article for publication, find out the precise 'house style' that is expected and stick to it — you will save yourself unnecessary revision and possible rejection in the long run.

- Be sure to provide a concise abstract that states the reason for the work, name of the client, main results and conclusions. This may be all the managing director or chief engineer reads; others will do the detailed assessment.

- Plan the report structure carefully.

- Vary sentence length. If all sentences are the same length, the result is boring, and a report that is all short sentences has an aggressive tone. A series of long sentences is even worse!

- Keep jargon to a minimum.

- Include illustrations if relevant.

- Conclusions should be clear and specific.

- Give proper acknowledgements and provide a bibliography with papers.

- Finally, before you write that paper, make sure it is necessary (don't write just to improve your publication average).

These guidelines also apply to technical proposals and case justifications. Try to put yourself in the place of the client or manager and then answer 'their' questions. Clients need to know what is in the proposal for them and their company. Make it clear!

When you have finished the first draft of your report, article or paper, go back over it and check methodically for errors and to ensure it makes sense. Then go over it again.

When you are sure you are happy with the final version, put your work away in a drawer and don't look at it for at least a week, or preferably more — you will be amazed how many further errors and opportunities for better expression jump off the page. Keep a separate note of any ideas for improvement that occur to you in the meantime.

Finally, practise. People who write for a living know that there is no ideal way to say something. Even top authors are constantly refining their style.

Top tips – the written word

✔ State clearly why you are writing a letter or memo

✔ Find out your organization's house style and stick to it

✔ Always use a spell-check — but don't trust it too far!

✔ Read and check written communications before sending them

3. Telephone skills

Complex matters that require detailed discussion merit an email or letter, not least because they form a written record of what's been discussed or suggested. But if your question or message is reasonably brief, use the phone. Talking to someone directly can help forge personal links and reduce the chance of misunderstanding.

Cultivate a personal, friendly style. A good telephone manner is a great asset. Continue the call until a satisfactory and clear understanding has been reached on both sides, particularly if you are dealing with a complaint.

General tips

- Always give your name and company name when making a call.
- Never just say 'hello' when answering the phone; give your department and your name.
- State your business clearly; people are rarely sitting waiting for you to ring.
- Make sure you know who you are speaking to and what their position is.
- Take notes as you speak and listen.
- Relax and use an interested tone of voice. If you are having a bad day, don't pass your grumpiness on to the other party.
- Smiling while talking on the phone will give an impression of friendliness.
- Be reasonably brief without being curt.
- If you do not know the answer, say so, and offer help in finding it.

- Say you will ring back if you need time to think. Don't make hasty judgments.
- Do not allow a call to interrupt a meeting or interview unless it is vital.
- Make it clear that you are following the speaker's argument; they cannot see your face, so some verbal encouragement is needed.
- If you are faced with a complex message for a colleague, either say you will get them to ring back or suggest the caller should write a note.
- Whatever you promise to do as a result of a call, do it promptly!

Dealing with telephone enquiries

Remember that an enquiry could turn into an order, so deal with the caller efficiently and courteously. Write down:

- Name of enquirer.
- Position.
- Company address and telephone number, plus fax number if they have it.
- Company business.
- Nature of enquiry.
- Time you took the enquiry and the date.

Then:

- Record what you have done; for example, 'I will get X, the engineer in charge, to ring you back today' (check with the caller if they will be available).

- Make copies for those involved in the response, and keep a copy for yourself.

- If need be, check that action has been taken.

Your company may have a telephone enquiry system. If not, you could suggest it is worth having both a system and a follow-up procedure.

Finally, you may speak to unsolicited callers who ask for information but whose purpose is not clearly stated. If it is not clear why they are calling, check:

- What information is needed and why they want it.

- Who is speaking and what their company's business is.

Make it clear that they should make an appointment to see the appropriate person, or ask them to put their request in writing to the company secretary. Do not agree to see them yourself, or answer a few questions — you could waste hours!

Dealing with irate callers

- Irate callers are often incoherent; rage and frustration can render them incapable of even telling you their names, let alone the reason for their call.

- Try to get the caller to tell you the story from the beginning: say, 'I will try to help you but I do not have the whole story, please tell me'.

- Do not offer to transfer them until you are sure you are putting them on to the right person — the wrong one will enrage them further!
- Be helpful and calm.
- Get them to give you their name, address and phone number, and the cause of their complaint or problem. Promise to ring back when you have found out what is wrong, or get the right person to do this. *Then do it!* Don't think that getting them to put the phone down lets you off the hook.

Voicemail

Voicemail works like a telephone answering machine. If you are unable to answer a call, the system will automatically prompt the caller to leave you a message. When you next return to your telephone, you can play back the recorded messages. Most systems offer the following features:

- Message waiting — the system tells you of any new messages.
- Remote access — you can access your messages from any telephone.
- Forwarding — messages can be transferred to other voicemail subscribers.
- Password protection — you can control who has access to your mailbox.

Respond to your voicemail messages promptly. This will encourage people to leave messages in the future. If you are unable to return calls for an extended period, customize your greeting accordingly and give alternative numbers that callers can try for assistance.

Create your own personal greeting and update it regularly. When callers hear a current personalized greeting they are more likely to leave messages. The customized greeting should indicate the date and time the message was recorded, and a likely return time.

Offer an alternative number and ensure the recipient of these calls is available to answer their telephone and that they are able to assist callers with their enquiries.

Don't be tempted to use the call-forward facility to avoid answering the telephone. Callers will appreciate the voice message option but they would rather speak directly to you.

Mobiles

A mobile phone gives you the freedom to call from wherever you are, providing the area is serviced by a network. Many models also offer other facilities, such as a contact book and options for accessing the web. They may combine the functions of mobile phone and basic word processor, or PDA (Personal Digital Assistant).

If you want to take your office with you, specialist software can be obtained that allows you to link your laptop computer to your mobile, so making it possible to send and receive email, and access the web from remote locations. The same software may also allow you to synchronize the files on your laptop or PC with the address book and database in your mobile.

Keep calls reasonably short — there is no conclusive scientific proof that mobile phones cause health problems, but it is prudent to avoid using your mobile for extended lengths of time.

Don't use a mobile while driving unless the vehicle is fitted with a hands-free kit. Doing so is a driving hazard and is also illegal in many countries. If you need to use a mobile while on a journey, pull over and stop before doing so.

If you are going to use your mobile abroad, check before you go that it has been activated to work on foreign networks.

Lock keypads to avoid mobiles making accidental calls

Make sure you keep your battery charged, and that you have a mains or car battery charger with you. It is useful to carry a spare, fully charged battery just in case you need it.

Turn off your mobile before going into a meeting, media interview or other situation where an unexpected call might cause disruption or embarrassment. Set your mobile to 'vibrate' mode if you want to be alerted to incoming calls.

Mobiles left loose in a pocket or briefcase can easily be activated accidentally and call numbers stored in your address book. Lock your keypad to avoid embarrassment and a big bill — or worse!

Mobile phones can pose a safety hazard. Do not operate on filling station forecourts, aircraft in flight or near chemical plant or technical equipment that may be sensitive to the phone's emissions.

Mobiles and confidentiality

Avoid discussing sensitive or confidential matters on a mobile phone if you are in a public place, especially bars, restaurants and trains. It is all too easy to get engrossed in a conversation and forget that everyone within earshot can hear you. You never know who is listening in!

Less likely, but still possible, is the chance of your phone conversation being intercepted electronically.

Texting

Many mobiles allow you to use the keypad to send a short text message to another mobile or a computer. This can be especially useful when speech is not possible, for instance if you are in a conference or a meeting, or the person you are sending to is not in a position to answer their mobile.

Typing in a message can be a fiddly task. Fortunately more sophisticated mobiles are able to anticipate the word you intend to type from the context of the message, so saving you the effort of laboriously entering every letter of each word.

Even so, screens and keypads on most mobiles are small and awkward to see and use, so texting is only really practical for short messages: a brief question, query or comment.

The need for brevity in text messages has generated a cut-down form of English. For example, 'I will see you at the meeting at 6 tonight. John', can be entered onto your keypad as 'CUmtg6.J'.

Only use simple abbreviations that you are sure the receiver will understand. Do not invent long acronyms that may seem clever to you but are obscure to others. Review your message before you send it and ask yourself if the meaning is clear.

Top tips – telephone skills

✔ Be polite and friendly

✔ Say you will ring back if you need time to think

✔ Find out who callers are before committing yourself

✔ Be careful when discussing confidential issues using a mobile

✔ Always have a charged battery with you

✔ Use text to communicate when speech is awkward or impossible

4. E-communication

Electronic communication has revolutionized the way we talk to one another. Email, video-conferencing, online meetings and so on make it possible for us to exchange and discuss information around the globe without us travelling a step from our desks.

Email

- Only send email when you have a definite purpose. The more you say, and the more often you say it, the less likely it is anybody will pay attention.
- Always include a subject line in your message. This will allow the reader to prioritize, search, and file the message easily. For example, when sending a message about a meeting, the subject line 'Production Meeting Thurs 15th, 1000' is more meaningful than just 'Meeting'.
- Keep messages concise and to the point. Put important information in the first paragraph so the reader can immediately decide its urgency and relevance. Make your message easy to digest by using bullets and paragraph breaks. Paragraphs should be no more than three to four lines in length.
- Check and read through emails before sending them. Poor spelling, major grammatical errors and 'typos' or mistakes will give the impression of sloppy message management. Use a spell-checker. But don't rely on the software too far. The best spell-checker in the world can only check spelling, not 'weather' you have chosen the 'write' word — as this sentence demonstrates!
- Limit the use of distribution lists, as the information in your message may not be useful to everyone. Be careful when replying to distribution list messages; ensure that you reply only to the sender and not to the entire list, unless a reply to all is appropriate.

- Always include your name and alternative contact information in your message. This allows people to call or fax you information more easily.

- Check the download time of attached files. Large PDF files, pictures, audio and video clips can be several megabytes in size. If the recipient has a slow internet connection, downloading your message may take many minutes, clogging their system and causing irritation. If you must send a large file, contact the party you are sending it to and warn them beforehand.

- If you are away from your desk for any length of time, set up an AutoReply message. Otherwise those messaging you may think you are either too busy or uninterested to reply.

- Remember that an email is like a postcard — potentially anyone can read it. Include only content appropriate to your professional position when at work. Also, an email is its own record, which could be held against you.

- Never send emails in anger, and be aware that it is easy to give the wrong impression with an email. For example, you may come across as more abrupt than you might intend. Sometimes a face-to-face meeting or telephone call can be the best means of smoothing over a difficult situation.

- Some email systems allow you to recall emails that have not yet been opened, but in general you should assume that once an email has left you it will be seen. Always think through the potential consequences of email before you send it.

E-tiquette

Email is a halfway house between written and verbal communication. One big difference is that email can fail to convey the subtleties of face-to-face meetings. The following suggestions can help to save embarrassment and misunderstanding:

- DON'T SHOUT! Using capital letters throughout your message is the equivalent of yelling at someone. It is also more difficult to read. Mix upper and lower case (as here); it's more friendly and much easier to digest.
- Don't put a long list of addressees in the To: or Cc: fields. Each addressee will be able to see who else is on your distribution list, which might not be a good idea. Use the blind copy field (Bcc:) instead. This way the privacy of everyone on your distribution list is preserved.
- Think twice before forwarding rumours, chain mails, jokes or unsubstantiated virus alerts. The recipient may not share your sense of humour, and false virus alerts can lead companies or individuals to take unnecessary and time-consuming counter measures.
- Don't send anything confidential. Email is not secure, and it only needs a simple typing error or technical glitch for your secrets to end up in the wrong hands.
- The normal rules of polite conversation apply. The word 'Please' can work wonders, as can 'Thanks'.

Emoticons

As noted above, emails (and text messages) can be taken the wrong way, especially if irony or humour is intended. The recipient cannot see or hear you and so may take your message at face value. To get around this problem, computer users have developed sets of punctuation marks called 'emoticons' that can help put your message into its proper context. For example, a smile :-) added to a message can draw the sting of what might otherwise be regarded as rude or brusque content.

Other useful emoticons include:

:-o Surprise

:-(Frown

;-) Wink, or tongue-in-cheek

Although useful in informal communication, emoticons may be viewed as frivolous, so their use is probably best avoided in serious or formal situations.

Online meetings

Online meetings offer you the chance to communicate with other people using the internet, your organization's internal intranet or by modem. Tasks such as setting up a meeting, inviting others to take part and actually holding the meeting itself are made quick and easy by your computer software. This may be proprietary software designed and created by your own IT department, or commercially available programmes such as Microsoft Netmeeting®, which integrates with the Calendar and Address Book in Microsoft Outlook®.

Whatever your software, an online meeting will include the facilities to:

- Send and respond to messages in real time.
- Speak to other people in the meeting.
- Share files, video and audio.
- Work together on the same file.
- Illustrate ideas on a communal whiteboard.

Procedures for setting up or taking part in a meeting will depend on the exact software you are using, but will generally follow the same pattern:

- Schedule the meeting.
- Invite other participants.
- Log on to the meeting.

When taking part in an online meeting:

- Keep messages precise and to the point.
- Check and read through messages before sending them.
- When replying to messages, do not send a new message with a new subject line; it can be confusing. Stick to message threads that connect a series of related comments and replies.
- Don't assume people will know what you're talking about or are replying to.
- Check the download time of attached files.

User groups

There are a huge number of user groups available on the web covering diverse and sometimes very specific personal interests. Professional institutions are seeing the value of setting up this type of facility for their members to share technical or professional ideas and opinions. Examples include the Institution of Chemical Engineers' e-networking toolkit (http://www.icheme.org/enetwork/) and the Institution of Electrical Engineers' Professional Networks (http://www.iee.org/).

A user group is a collection of computer users who share responsibilities and perform similar tasks, or who are involved with a specific project. Members are able to post and reply to messages, so creating a form of virtual meeting or discussion forum. User groups can be single tier or multi-tier, allowing access to a range of different individuals linked to a single project or task.

The following tips should help:

- Make sure the subject line is completed, and be specific. Avoid vague subject line entries.
- Only include relevant return text when replying.
- Use the spell-checker, and proof-read messages before sending them.
- Make sure your system is virus-protected. If you inadvertently admit a virus to the group you could be barred from further participation until you have disinfected your system.

If you are the administrator you will be empowered to create and modify the structure of a group and establish different kinds of security and access levels.

Video-conferencing

A well run video-conference can offer an effective alternative to a traditional physical meeting, but success does require careful organization, especially if the conference involves several participants at different sites (a multi-point conference).

- Be on time. Other participants will not appreciate it if they are ready on time and you are not.
- Check the camera can see you properly, and also that you appear level with other participants, not looking down from above or peering upwards.
- Adjust lighting levels to give the best possible brightness and contrast. Check for visual distractions behind you and if possible remove them. Ensure microphones are positioned to give good sound quality, and run a sound check before the conference begins.

Facilities should be similar at all video-conference points

- Maintain eye contact. Video-conferencing is more like a personal exchange than a telephone call, so body language and facial expression are important. Doing other things or looking elsewhere has the same effect as ignoring a person during a face-to-face meeting.

- Don't hold side conversations, and keep distracting sounds to a minimum. The non-selective nature of microphones can make it difficult to distinguish the primary discussion. Extraneous noises and general chatter can complicate multi-point conferences in particular.

- Ensure that everyone participating in the conference has access to the same supporting materials and resources. This may require prior planning if hard copy or plans and drawings need to be sent to every node of a multi-point conference.

- If you are organizing the conference and have invited outsiders to a point away from their base, arrange for them to have access to refreshments, etc. If the standard of hospitality varies considerably between sites, try not to make it obvious!

Websites

Developing a website, either for an external audience (the internet) or for internal use (intranet), can be a very effective means of communicating key messages to clients and staff.

Creative web design is a skill, if not an art, but we can all follow some basic rules that will help ensure the results are attractive and readable:

- Draft out what you want to achieve on paper first. If the site is to include several pages, think through how they are going to work together.

- Be sparing with jazzy patterns; they will distract and even irritate the user. Simple or plain backgrounds work best.

- Don't be afraid to include lots of information on each page, as users will prefer this to having to click through to different pages. BUT ensure that the page doesn't end up as an unreadable mess! Clump related information in one area of the page, and consider using different coloured text or colour blocks to help users find their way around.
- Avoid boldly coloured backgrounds, and especially very dark colours. Again, they can distract or irritate. White or plain pastel shades are preferable.
- Don't be too clever. Users will prefer a simple site that downloads quickly. Intricate links, hidden hotspots and special effects work well, but only if they do not make the page unmanageably large or confusing to navigate.
- If you plan to embed photographs on the site, cut them down to size. Users with slow modem connections will not wait for several minutes for a 'data intensive' image to download. As a rule of thumb, an image around 300 pixels wide should be no bigger than 15 KB in size.

Top tips – e-communication

✔ Check emails before sending them

✔ Only send emails (or copy mail) when there's a point

✔ Don't send anything you regret — it may come back to haunt you!

✔ Observe email etiquette

✔ Keep websites simple and easy to use

5. Body language

What we say is only a part of communication. Over half the information we get from other people comes from non-verbal language. The way we dress and our gestures and expressions are just a few of the complex range of non-verbal communication tools we use, often unconsciously, to express ourselves. As a result we often give other people more information about ourselves than we may want to reveal.

Knowing how body language works and controlling it can be a powerful communication tool — helping you to influence other people, make a sale, get a job or win promotion.

Dress sense

The clothes you wear say a lot about you. Dress for the occasion and according to the conventions of your organization. While jeans and T-shirt may be appropriate in some companies, a 'collar and tie' will be more suitable in others, especially on formal occasions when you are trying to give a positive and professional impression, such as client meetings or job interviews.

A neat well-groomed appearance need not brand you as a 'suit'. What it will do is indicate to others that you respect them and the situation. It may also give the impression that you can control and adapt your personal environment to circumstances and, by extension, will be able to do the same in your work.

Safety or other appropriate work clothing may at times be required for operational reasons and such practical dress codes should be observed.

Expressions

Smiling is a universal means of expressing friendliness. But a forced smile can look false, and overdoing it can give the impression you are hiding your real feelings, or it can make the other person feel patronized.

Non-verbal signals are an important part of communication

Scowling expresses disapproval. Try and limit the scowl to circumstances that justify it. Many people, men in particular, unconsciously adopt an over-serious or even truculent mien as a matter of habit, giving an erroneous impression of bad temper or even anger.

Firm eye contact is a very important aspect of body language. It suggests you are truthful, transparent, honest, sincere, confident and in control of the situation. On the other hand, an unwillingness to meet the other person's eyes implies disinterest, contempt, shiftiness, or that you are hiding something.

Having said that, an unblinking stare can be unnerving as it suggests dislike, contempt or even aggression.

If you want to convince someone of your businesslike approach, focus your gaze on an imaginary triangle in the lower centre of the forehead, just above the top of the nose.

Mannerisms and gestures

While some mannerisms and gestures are unique to us as individuals, many are sufficiently universal for others to be able to interpret them. Knowing what to look for can give you an insight into how other people are thinking, which can be very useful at critical meetings. But do be aware that others may be monitoring you in the same way!

Watch out for the following:

- Biting the nails and / or pacing up and down can telegraph nervousness or uncertainty.

- Licking the lips in conversation can suggest the person is nervous or has just told an untruth.

- Touching the face, scratching the nose, looking down and to the left or a hand over the mouth can also mean the person is not telling the whole truth.

- Crossing the arms or legs is a 'closed' or defensive signal.

- Leaning forward when seated at a meeting or interview can imply attentiveness and interest. Leaning back creates distance and implies disagreement or disinterest.
- Aligning your torso with that of the other person implies liking, agreement and loyalty. Turning away is an independence or power gesture.
- Foot-tapping, doodling and gazing into space indicates boredom or impatience.
- Repeating other people's positive body language will signal that you like them. Mirroring negative signals will have the opposite effect.
- Limp handshakes are out. A good firm handshake implies confidence. But don't go to the opposite extreme and crush your opposite number's fingers!
- Slinking, or conversely, rushing into a meeting or interview room can imply lack of confidence — pause for a moment at the door and then enter at a normal pace.

Gestures and non-verbal language can be a particular problem when dealing with other cultures. For more on this see Chapter 12 on working with global teams.

Top tips – body language

✔ Dress for the environment you are in

✔ Smile — but don't overdo it

✔ Good firm eye contact suggests sincerity

✔ A firm handshake implies confidence

6. Job applications and CVs

Whether you are looking for your first job or trying to improve your career prospects, can communication skills do anything to enhance your chances in the current job market? The answer is 'Yes'.

Your job-hunting could begin by:

- Studying vacancies advertised in the press.
- Making a list of companies you would like to work for.

Alternatively, you could contact selection consultants and agencies. The very fortunate might obtain a personal recommendation or be approached by a 'head-hunter'.

In times when jobs are hard to come by, advertised vacancies attract hundreds of replies. Writing to companies you'd like to work for, but who are not advertising, can be more rewarding as your details will be on file when a vacancy does come up.

Try to be both systematic and realistic about your approach. Ask yourself:

- What sort of job do I hope to get?
- What characteristics should the holder of that job possess?
- Do these characteristics match my own?
- Does the job suit my qualifications and temperament?

Do your homework before you tackle the actual application. Note any mention of the companies in the business pages of the quality newspapers and specialist technical press. A day in the reference section of your local library will not be wasted.

If you are answering a job advert, study it in relation to what you have learned about both the company and yourself (this is equally appropriate if you are just writing in hope).

Have you uncovered any clues that will help you in the preparation of your CV, and the covering letter you need to write?

If you have read and acted on the other advice in this book, you should be able to write a clear, confident and well-expressed letter, and if you need to telephone, your manner will be friendly and pleasant. However, a few extra tips may help to make your covering letter work for you.

The covering letter

Your covering letter should be polite and reasonably short. Yours will be just one of tens, even hundreds, of similar applications. Make it easy for the recipient to read:

- State clearly the reason for writing.
- Give a reference — either use the exact job description (copied from the advertisement) and say where and when you saw it, or indicate you are writing to ask if there are any suitable vacancies in the company.
- Highlight your major strengths and experience in relation to the job.
- Add a short sentence or two about your interest in or suitability for the job.
- Make it clear that you have enclosed a CV or application form, and that you hope to hear from the company shortly.

If your handwriting is bad, type the letter. Check the spelling as many personnel departments put misspelled or illegible letters on the reject pile straight away. Use good quality writing paper —

ruled paper from a scribbling pad makes it clear that you have made no effort at all.

Remember the letter is all the company has to go on in the first instance. *It is you!*

The CV

Limit your CV to two pages and staple or clip them together. The interviewing panel or person may have to read dozens of CVs, so make yours the easiest to read and handle. Remember that the interviewers may have to read all this material after they have finished their usual work. Make sure they don't give up halfway through. Keep your application short, neat and easy to handle.

Summarize relevant experience. Although the facts about you are constant, you may need to alter the emphasis you give to them, depending on the job you are applying for. If you have studied yourself and the company objectively, you will realize what adjustments you need to make. Just as each covering letter will be different, so your CV should be addressed to the specific job.

Add details of your education. Put your degrees first, and just your highest qualifications at school. Note that employers can run a check on degree status, so think twice before awarding yourself an honorary doctorate! As a subsection you can list other qualifications, but do not list every short course you have ever attended.

Outline your career so far by listing your employment history (in reverse order). Most employers will check references so, once again, take care you can substantiate your claims.

Include limited personal information, such as interests. Keep it short and specific: for example, not 'sport' but 'basketball'.

Application forms

Of course, all this effort may result in no more than an application form. Keep a copy of your letter and the CV as sent to each company, and use the same information so that there are no discrepancies. It saves time too!

Educational and professional qualifications
may be checked

Forms may be provided in computer format (e.g., Microsoft Word®) or on the web, in which case filling in the spaces provided is straightforward. Otherwise complete the form in neat handwriting. Do not use capital letters throughout; they are difficult to read and you may run out of space, making the last line impossible to decipher.

Return the form with a covering letter reiterating (in different words) your interest / suitability for the job. If there is a named person on the advertisement or covering note with the form, write to them 'Dear Mr/s Blank' and sign yourself 'Yours sincerely'. Use 'Yours faithfully' if you have to write 'Dear Sir / Madam'. Keep copies of these items too.

Finally, be patient. If you do not hear for some time, you could phone the personnel department and ask if there is any news of your application. Be prepared, not only with the facts and figures — for example, date of the job advertisement and when you last wrote — but also for possible disappointment. In such cases, be polite and friendly, and start again.

Electronic CVs

Increasingly, larger recruiters are using computers to sieve through CVs for suitable applicants. CVs received are stored in databases. If your CV is to stand out you need to include a keyword section listing your skills. This increases your chances of being short-listed by the computer for more detailed consideration. An electronic CV will be filed automatically, while written CVs will be scanned in for later retrieval.

Make it easy for OCR (optical character recognition) software to read your CV:

- Use white paper and plain black text, and avoid colour, fancy borders or other decoration.
- Don't use columns — stick to simple layouts.
- Don't use small font sizes — 10 point should be the minimum.

Compile your CV using a common word-processing package, such as Microsoft Word®, that can be easily read by the recipient. You can then attach the Word® file to an email. If you are not sure, distribute your CV as an email in plain text, or 'ASCII', form.

Watching your back

Many companies monitor email traffic sent by staff. Be aware that your quest for another job could become known to your current employer, with awkward consequences. If possible send such communications from home.

Also, posting your CV to a website or sending it to a site that can be accessed by others has its own pitfalls. Your details may be used to send you junk mail. Worse, if you include too much personal detail there is a danger someone may attempt to assume your identity!

Check:

- Is the site secure?
- Is it password protected?
- Who owns and runs the site?

Never include crucial personal information (driving licence number, National Insurance number or contract details / staff number).

Top tips – job applications and CVs

✔ Target your application at a specific job or company

✔ Make it easy for the recruiters

✔ Ensure both your CV and covering letter are concise and legible

✔ State clearly how your background fits the post advertised

✔ Before sending, check your application and CV for errors

✔ Don't post confidential private information to public websites

7. Job interviews

While a CV contains details of your skills and experience, the interview will give selectors a better idea of whether you as a person and as a professional will fit the company culture. Prepare with this point in mind.

The days before

- Find out as much as you can about the company and the post you are applying for. Contact the employer and request a copy of their annual report and corporate promotional literature. See if you can find out information about the job, or at least the type of job. Check out relevant websites.
- Go through all the correspondence with the company; study your copy of the job application and the material you collected on the company's operations.
- Prepare questions that will make you appear interested in the company and your possible future with it.
- Consider your appearance. The watchword is professional: clean, tidy and well turned out. Your shoes may be old but they should be clean!

On the day

- Make sure you leave plenty of time to get to the interview, and allow for travel delays, difficulty in parking and so forth. Aim to arrive about 15–20 minutes before the stated time; it's better to be early than risk being late. Take the opportunity to visit the bathroom if you need to.
- When you are met by the personnel officer or secretary, find out the name of the principal interviewer and his or her position in the company.

- Remember that the people you meet before you go into the interview may well be asked for their opinion after you have gone, so be polite and pleasant to all.

- When you go into the interview room, keep calm and don't rush. Smile, greet the interviewer or panel, and shake hands if this is expected. Sit when invited, sit still and look alert.

The interview

- Listen carefully to the opening remarks, make eye contact, and study the body language. When answering questions keep to the point, but avoid abrupt, monosyllabic replies like 'Yes' and 'No'.

- Concentrate on your strengths when the opportunity occurs. Be positive, optimistic and friendly in your replies and avoid any suggestion of inflexibility. Ask for clarification if in doubt.

- Remember that the interview is your opportunity to show yourself at your best (and that does not mean showing off). It is also an opportunity to learn about the organization you hope to join. When you can, ask the questions you have prepared on company organization / training / experience in other departments or other matters. Do your bit towards keeping the dialogue going.

- If the interviewer just says 'Tell me about yourself', say 'In relation to this job I believe I have the following strengths …'. This may lead them to invite you to describe your personality, or your interests. Do this briefly but always bear in mind how you wish to present yourself.

- Use humour if you feel it is appropriate, but beware, they may not have a sense of humour!

- If you find the interviewer is aggressive, keep cool, maintain your principal points but concede gracefully where you can. If they push you on a particular issue, try turning the question back: 'Could you explain why this factor / point / procedure is so important, please?'
- Silence may be another ploy, in which case ask an appropriate question from the list you have prepared. If you are the last candidate in a full day, they may be very grateful, as the silence could just be interview fatigue.

Some final thoughts

Most interviews come down to the question, 'Will this person fit in with our organization and get on with the people who already work here?' Highlight those strengths you have which would help you to fit in.

Your education and previous experience will have been studied and will have influenced your position on the short-list. These factors are important but friendliness, adaptability, a generous outlook, a willingness to co-operate and other character attributes will be equally important.

Cool judgement, firmness under pressure and skill in handling tricky situations may also be required. If you are a good communicator, your chances are improved.

Top tips – job interviews

✔ Do your homework beforehand

✔ Have a list of questions you want to ask about the employer

✔ Concentrate on making a good impression

✔ Give concise but comprehensive answers

✔ Stay cool and stand up for yourself if challenged

8. Meetings

You will have to attend briefing meetings and committees of some kind during your career, probably sooner rather than later. Meetings involve knowing the correct procedure to follow, basic psychology and not a little internal politics. To be effective they also involve good communication skills.

With a good chair, an efficient secretary and a carefully selected group of members, committees can bring the benefits of wide experience, a range of technical and professional expertise and the democratic process to effective decision-making. Badly run committee meetings can be costly, time-consuming and very boring.

Briefing meetings are essentially a method of communicating with the work team. You will want to find out about progress, problems, snags and so forth. Meetings can also be used to check how well the team understands management decisions and overall company goals.

Committees

Meetings can be intimidating at first, especially if you are new to the organization. You may feel everyone is looking at you, or sizing you up. In reality, they will probably be far more concerned with their own brief, so relax. You may find the following tips helpful:

- Prepare well — if you have been given a specific brief, make sure you do your homework and have something to say when the time comes.
- Follow the chairperson's lead and address all remarks through the chair.
- Listen and watch at first; no-one likes the newcomer who keeps on interrupting and putting themselves forward. Get a feeling for who's who on the committee and how they relate to each other.

- When you do speak, make your points calmly and succinctly.
- Expect to have to justify your arguments. Listen carefully to questions or counter-arguments put forward before answering them

Taking the chair

At some stage you may find yourself chairing a meeting. This can be quite a responsibility and will certainly involve you in extra work if the meeting is to run smoothly.

- Once you have been made chair, ensure that everyone knows why they have been invited to attend the meeting. If you can influence numbers, keep the committee small.
- Prepare the agenda well in advance. Check the minutes of the last meeting. Keep your objectives in mind.
- If you can, keep notes and write the minutes yourself. It may save much time and effort.
- Be impartial. Avoid favouritism and stop time-wasters or self-publicists firmly.
- Do not dominate the proceedings; a well-timed intervention is more effective.
- Be courteous to all, and ensure that everyone follows your lead.
- Summarize and confirm important decisions and allocate responsibilities clearly for each item on the agenda before tackling the next.
- Conclude the meeting by setting the date for the next meeting. It gives all present the opportunity to check their diaries.

Communication skills for secretaries

Traditionally the secretary remains silent at the meeting, other than helping on matters of fact and the minutes (nowadays the secretary may also be a well-informed participant). Procedure is also your responsibility so be sure you know the rules, and remind the chair politely if s/he strays.

- Consult the chair on the agenda well before the meeting. If there must be a change in arrangements, let everyone know as soon as possible. Phone first and confirm in writing.
- Ensure that the agenda and any other papers are distributed to all in good time.
- Don't try to record everything for the minutes. Write key words and phrases down. The chair's summaries of decisions and responsibilities are vital.
- Concentrate on recording ideas and discussion (not people).
- If you are not clear, ask for a point to be repeated or clarified.
- Resist pressure to 'improve' the minutes, but try to present contributions fairly and objectively (leave out the heated argument touches).
- Demonstrate your infinite patience and tact by reconciling the competing and inconsistent demands of the committee members!

Briefing meetings

Briefing meetings can be applied at many levels: from the managing director to departmental managers; managers to department or section heads; section heads or project leaders to their teams, and so on. They should be informal but, if you are doing the briefing, you should structure your remarks.

The briefing explains management or project decisions to the people who must apply them and make them work. You can cover the ground if you concentrate on the five Ps:

- Policy.
- Progress.
- Problems.
- People.
- Points for action.

Allow time for reaction as new problems and misunderstandings can surface. As the responsible person you must observe and listen, as well as talk, and you must be ready to resolve or explain unclear issues or points. If necessary, you must be prepared to take the matter up with your manager. Properly used, the briefing can help to weld the team together, giving members a sense of the importance of their contributions and a feeling that they count.

Once you have set up, say, monthly briefings, you should be able to ensure that they last about half an hour (one hour maximum) without seeming to curtail the exchange you are aiming for. The time should only be extended if it's obvious that there is an important misunderstanding that affects everyone.

If just one person has a problem, invite them to discuss things with you after the meeting: 'Hang on for a moment, X, and we'll go through that again just to make sure I have got it right'. That way the rest of the team can get on with the job, and you can resolve the issue quietly.

Top tips – meetings

✔ Prepare well

✔ Listen and watch before speaking out

✔ Be impartial

✔ Isolate problems before they get out of hand

9. Handling conflict

Conflict can arise in the workplace for any number of reasons, not least through clashes of personality or because strong-minded individuals can have private agendas.

You will find your job easier if you put effort into communicating with colleagues and other staff. You don't have to like them, but keep emotion out of your work, if you can.

The answer is not to take things personally. An argument is not usually an attack on *you*, and *your* disagreements with your colleagues should not be on a personal basis either. Remind yourself you are debating a point, whether of policy, technical feasibility or economic reality.

Many conflicts stem from misunderstanding. For example:

- Not listening properly.
- Careless reading of instructions.
- Poor instructions, with the most important detail left out, like the famous cooking recipe for jugged hare that begins 'First catch your hare'. Not as obvious as it seems!
- Lack of attention to the matter in hand by all concerned.
- Arguments about who should do the job.
- Making assumptions ('I thought you would do that') and not checking.

In spite of every precaution, problems will still arise. The potential for friction exists even in the best-run companies. For example:

- In technical situations colleagues can differ over options and solutions.
- Deadline pressures can upset some people, leading to delays and recriminations.

- Compromise over safety and economic considerations can cause morale problems.
- Collaborators may turn into rivals if promotion beckons.
- Competition may make the world go round but it also promotes conflict.

Building bridges

Bearing such factors in mind, you will have to accept that sometimes conflict is inevitable. You need to be alert, ready to note the unexpected reaction and, when necessary, be willing to admit that your argument has weak spots. This can take courage, but communication is all about dialogue. Being human once in a while will encourage others to relate (and therefore talk) to you.

- Ensure that a simple misunderstanding is not the cause of the trouble.
- Face up to a problem positively.
- Do not provoke; your objective should be to get the best deal worked out.
- Stay calm and be polite. Remember that you have to face your 'opponent' tomorrow, or next week. While both of you work for the same company, try to keep a friendly tone and a reasonable manner.
- Give way gracefully yourself when the argument goes against you and be willing to accede defeat. But don't give in too quickly just to keep the peace — after all, you could be right after all!
- Listen to others; there may be a face-saving formula for both sides that leads to agreement. Be prepared to agree to differ.
- If debate is justified, keep cool and review the pros and cons to help isolate the real problem.

- Don't look for trouble; a few checks might enable you to put the matter right quietly and efficiently.
- If things look bad, suggest a cooling-off period, or sleep on the problem.
- Make sure you know what company policy is before you start banging your head against it.
- Remember that your 'opponent' may not have the authority to change the party line.
- Leave the 'opposition' an acceptable escape route and don't rub their nose in it if you win.
- Ensure you survive to fight another day — after all, you can only resign once!

Handling a hostile audience

There may be occasions when you have to face a hostile audience. For example, this could occur if you are attending a public meeting as part of a public consultation about a controversial project or development.

Susan Escott, a UK based expert on advanced presentation skills, suggests the following points on communicating effectively with a critical or angry audience:

- Think through what your audience may be expecting from the meeting.
- Prepare answers to obvious questions and points.
- Remain calm.
- Be polite, professional yet pleasantly assertive.
- Confront speculation.

- Allow people to 'have their say'.
- Empathize — it's difficult to argue with someone who agrees and / or shows sympathy.

For more on the linked issue of public attitudes to science, see Chapter 13.

Top tips – handling conflict

✔ Stay polite

✔ Listen to the other point of view

✔ Don't assume you are right, but you could be, so don't be too defensive!

✔ Don't 'burn your bridges'

✔ Empathize

✔ Don't take things personally

10. Public speaking

Whether you are holding a conversation or addressing a large meeting, you need to remember that your listeners are not just taking in the words. They are also registering your tone of voice, your body language and your general manner and appearance.

Conversations and visits

Getting out and about and meeting people — staff or clients — sets up good communication channels. Networking is the lubricant of commerce and many important orders arise from good personal contacts. Potential staff problems can be resolved at an early stage, simply because people feel they have access to you. Whatever the nature of the meeting, the following suggestions should ensure your reputation as a diplomat:

- When you greet a visitor, look them in the eye, smile and sound pleased to see them.
- Small talk is never a waste of time. Half the battle is getting started. Prepare a few general enquiries about the journey or the weather.
- If your memory is bad, keep notes on people you deal with regularly so that you can ask about their promotion / family / new car and so on.
- Know when to move the conversation on. Body language should tell you — your guest's if not your own!
- Listen carefully throughout the day; the most useful information may be given to you in the car park as you see the visitor off or you leave.
- If the visit is over-running or going badly, have a few polite disengagement strategies ready, but use them subtly to avoid offence.

- Let your guest / host finish speaking; don't interrupt in your eagerness to make your point.
- Time is money, so try to keep to the point once the real discussion starts.
- If you are going to deliver a bombshell, do it tactfully.
- Don't laugh at your own jokes; they may not be quite as funny as you think.
- Finally, if you are relaxed, friendly and enthusiastic, your visitors will be too.

Speaking to an audience

Your public could consist of two or three colleagues, a group of clients, or the audience at a major technical conference. However large or small the numbers, you need to prepare your presentation carefully: the words, the visuals and the timing.

Preparation

- Think through the structure of the talk carefully and logically, listing key points or messages.
- A good speech or presentation should appear spontaneous, so do not read from a prepared script. That said, at the preparation stage it may help to write out a draft presentation, and then read it out loud to give you an idea of whether your intended approach works. Cut out anything you cannot say easily.
- If there's a time limit to your talk, read through your draft presentation against the clock. You may be surprised how much over or under you are on the timing, in which case you will have to add or cut material, as appropriate. As a rule of thumb, speakers usually run over rather than under time, so think twice before adding more material at the last minute.

Check the body language of the audience for signs of boredom

- Using the written draft, prepare cue cards that give the key words and phrases.
- List the visuals to be used with each card — use a different colour or give them a column on their own so that you do not overlook or muddle them.
- Prepare a summary of the main points of your talk, if appropriate, ready to hand out afterwards.

Timing and delivery

- After the initial greeting (for example, 'Good morning, ladies and gentlemen'), tell the audience what you are about to tell them. Once you have made your points, end by summarizing what you have told them. It will help them to remember your contribution.
- Be yourself. Use words you are comfortable with.
- Speak up and speak clearly. Don't gabble or pause distractedly (the audience will think you have forgotten what you were saying).
- Try to adopt a conversational approach. Put animation into your voice, **emphasize** key words and use pauses as punctuation, allowing your listeners to digest each point before you move on to the next.
- Humour is good, but be careful with set jokes. It takes years to develop a comedian's sense of timing.
- Stick to essentials; don't try to show how much you know or have done.
- Leave time for discussion. Plant a question or two in the audience's mind.
- Do not fiddle with your keys, loose change, or the desk furniture. It is very distracting. Use your hands to emphasize points naturally.

- Look at the audience. If making a presentation using audio-visual aids, do not face the screen and talk to it!
- Don't rush about too much. Pacing up and down before the audience will give them all 'tennis match' neck. Keep an eye on the audience's body language and adjust your approach accordingly.
- Keep to the time allowed. Know when to stop.

For more on creating audio-visual presentations, see Chapter 11.

Top tips – public speaking

Conversations and visits:

✔ Smile and appear pleased to meet people

✔ Listen as well as talk

✔ Keep notes about regular contacts

✔ Remember small talk can lead to bigger things

Speaking to an audience:

✔ Prepare thoroughly

✔ Check timings

✔ Don't read from a script

✔ Speak naturally and don't rush

✔ Leave time for questions

11. Audio-visual aids

They say one picture is worth a thousand words. But it all depends on the picture, and on the type of talk you are preparing.

There are a number of tools you can use to add impact to your talk, lecture or presentation:

- Computer-based presentation software.
- Overhead projection (OHP).
- Flip chart.
- Whiteboard.
- Video.

For technical talks in general, keep your aids simple. The more complex your battery of projectors, the more things can go wrong. You also run the risk that a visually enthralling presentation full of bells and whistles may actually distract the audience from your core message.

Creating a presentation

Commonly available software such as Microsoft Powerpoint® allows you to create and print colourful OHP slides onto acetate sheets, which can then be projected onto a wall or screen. Specialist diagrams or drawings can always be drawn onto plain acetate sheets using pens.

A more flexible alternative is to utilize the full power of Powerpoint® with a computer-based presentation. This lets you seamlessly embed relevant photographs, diagrams, video clips and even sound files, and allows you to animate each slide so that words can be made to appear from different directions and in the order you choose.

You can then present straight off a laptop computer, if its screen is big enough and you are only working with an individual or a small group. When working with larger groups you will probably prefer to project the presentation onto a wall or screen using an audio-visual projector linked to the laptop.

The latest video cameras, computers and editing software make the production of video and sound files a relatively easy task. For detailed instructions on how to achieve a full audio-visual presentation, consult your software's help files.

Maximizing impact

- Never overdo the visual aids. A slide saying 'Hello, I'm John of Company X' is hardly necessary, though a small company logo on each slide can help to reinforce your brand.
- Keep the visuals simple; a cluttered slide will confuse and irritate your audience.
- Use colour in your slides, but avoid orange and yellow; they do not show up well when projected. For text only, white or turquoise on dark blue is pleasant to look at and easy to read.
- If you need to print off handouts, select the appropriate options in your software. Choosing to print in grey scale will save the expense of using coloured ink, and the software will automatically render text so it prints out clearly as black or grey on white.
- Use a common colour theme or design throughout your presentation; it will appear more professional.
- If you are working to a set time limit for your presentation, rehearse and check timings. You may be surprised how much longer (or shorter) your draft effort is.

- Try to limit the word content of each slide to five or six key words or phrases. Use a reasonable size, blocky typeface that will enlarge well.

- Reveal one point at a time; your audience is more likely to focus on what you are saying. Computerized presentations can be adjusted to reveal each line of script, as you need it. You can achieve the same effect in an acetate based OHP presentation by masking the slide with a sheet of paper.

- Remember to update each section of your talk with the appropriate slide. It is all too easy to get carried away and find that the audience is still looking at the slide you were talking about some while before. This is especially important if you have set your computer to reveal one point at a time.

- Use colour photographs whenever possible, but check that they actually 'say' something and are not cluttered with extraneous detail.

- Keep video clips short (normally < 90sec) and to the point. If you are new to video work, read the camera's instruction book before you start filming; shaky, messy and out-of-focus footage will impress nobody. If the presentation is very important, consider using a professional camera crew to produce the quality product you need.

- Professional projectors give a bright image that is clearly visible under normal room lighting conditions. You should not therefore have to draw the blinds unless the sun is shining directly on the screen. Avoid projecting in complete darkness; you won't be able to see the computer or projector controls and your audience may well nod off!

- When using someone else's computer or projector, always check and test the set-up before you start the presentation to make sure everything works. Some projectors won't accept video clips, for example, or you may have teething problems with audio files.
- Put the computer into standby or switch the OHP projector off when not in use.
- Don't become so fascinated by your own presentation that you forget your audience. Your talk will have most impact if you look at the audience and let them look at the screen.

Diagrams

Avoid the temptation to try and use the diagram prepared for a technical report in your talk. It will undoubtedly be too detailed, like the examples shown on pages 67 and 68.

If you have a printing department or a photography section in your company, check the facilities they offer. There are various methods of making crisp, colourful overheads, slides, computer-generated colour reproductions and so on, using many different typefaces, to help you to put the message across. One word of warning: give support services reasonable notice if you want their help. Do not breeze in and ask for 20 slides by midday as a matter of course. They need time to produce good work too!

Original diagram

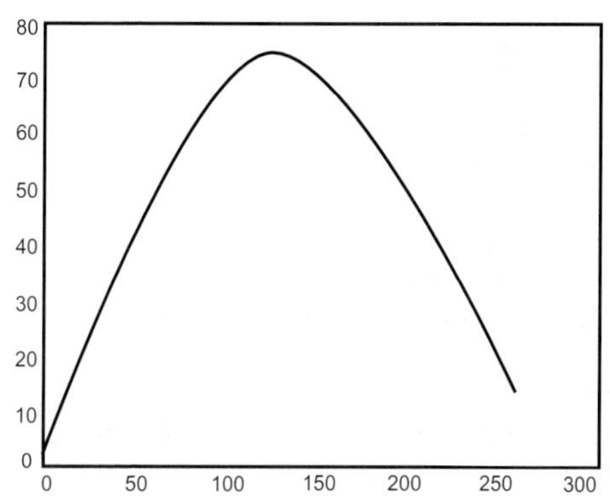

Fig 4 Example of savings in fuel costs when injecting fuel into a blast furnace

Visual for talk

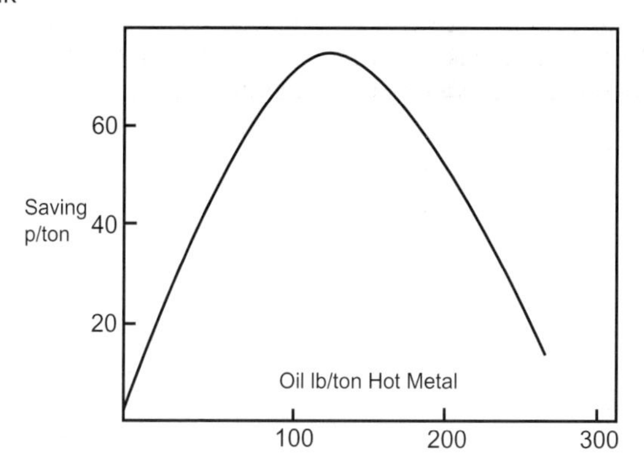

Original diagram

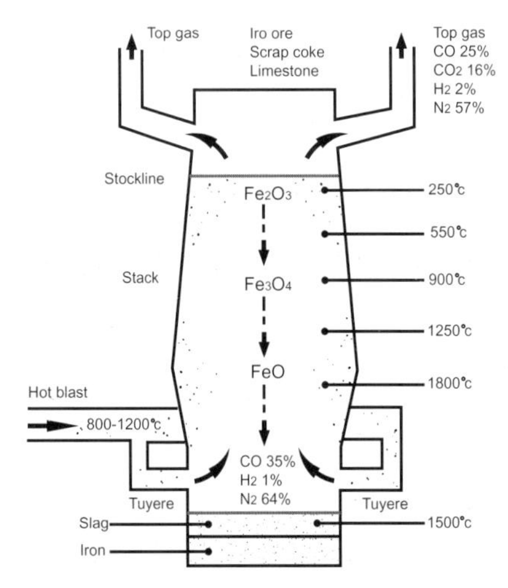

Fig 1 Principle of operation of a blast furnace

Visual for talk

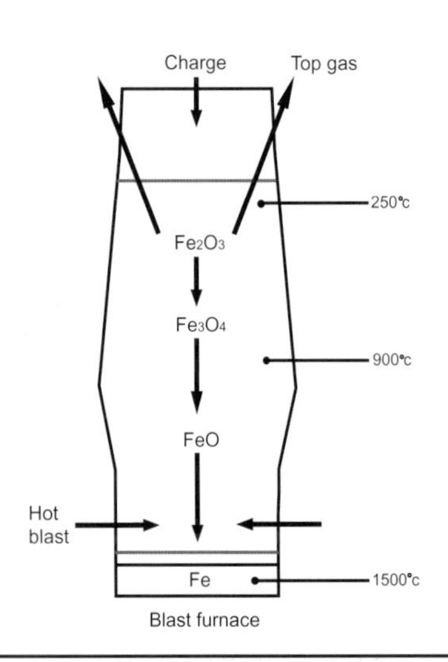

Blast furnace

Top tips – audio-visual aids

- ✔ Choose an effective theme and stick to it
- ✔ Keep slides simple, uncluttered and to the point
- ✔ Consider using sound and video to help put your message across
- ✔ Reveal and discuss one point at a time
- ✔ Rehearse and check timings
- ✔ Remember your audience

12. Working with global teams

Communicating across time zones

Increasingly, projects and the organizations that run them are multinational. You might be in London and have to talk to a colleague in Sydney, or be field based in Saudi Arabia and have to contact a client in London, Tokyo or Los Angeles.

Think about time zones before making calls. For example, when it's 10 o'clock in the morning in New York it could be midnight in Sydney, so don't be surprised if you get a somewhat sleepy, possibly abrupt, response.

Remember that some large countries can cover more than one, and sometimes many, time zones. For example, there is five hours difference between Hawaii and the eastern seaboard of the USA.

Unless the matter is urgent, try and keep calls to normal working hours in the country / time zone you are calling.

'It's only five inches on the map...'

If you have to liaise with personnel in a different part of the world, you will make your life and theirs easier if you remember that conditions vary enormously from country to country.

In some parts of the world communication links are rudimentary and prone to periodic disruption. Power supplies may be subject to frequent 'brown-outs'. Road and rail links can be primitive and slow. If your staff are not where you expect them to be, or you cannot get through by fax or phone, bear in mind this may be a problem well outside their control.

Some countries are huge by UK or European standards (though this will come as no surprise to those in North America!). A journey that may seem easy and straightforward on the map can in reality take colleagues several hours, even days, to complete. This, coupled with the problems of actually getting anywhere in some countries, can make it difficult for staff to meet deadlines that may seem reasonable to you.

Be sensitive to local conditions and make allowances.

Different folks, different strokes

Social customs vary widely between countries and cultures. This offers considerable potential for confusion or embarrassment when working with colleagues and counterparts from other parts of the world. By way of example, consider the different interpretations that can be placed on commonplace gestures and expressions:

- The familiar 'thumbs up' sign means 'ready' in the UK and USA, but is regarded as obscene in parts of the Middle East and Africa. Curiously, in Japan the same gesture signifies an order for five of something.
- Curling the forefinger and thumb together in a circle to indicate 'OK' in the UK and USA means 'zero' in France and will be interpreted as an insult in Italy, Denmark or Brazil.
- A shake of the head in India and parts of the Balkans means 'Yes', not 'No'. Rocking the head signifies 'Go on, I'm following what you say'.
- In Hong Kong and much of Asia it is offensive to beckon a person with a crooked finger. Instead, extend your hand palm down and flutter your fingers.

Expressions can also cause problems. For instance, most Asians think it rude to hold direct eye contact for any length of time, whereas on the other side of the Pacific an averted gaze implies shiftiness or disinterest. In the Far East, a smile following a mistake indicates deep embarrassment, not that the incident is being treated light-heartedly.

Cultures may interpret non-verbal signs differently

Work styles

Differences in attitude to work and business negotiation are widespread and can become a communication issue.

In North America and northern Europe rigorous planning and efficient use of time and resources are central to the work ethic. In southern Europe and much of the Middle and Far East plans may change easily, and a time commitment is often regarded as an approximate goal.

In the USA it is customary to be assertive, direct and decisive in business negotiations. In the Middle East and Asia business is only discussed after a lengthy exchange of social pleasantries and circuitous small talk.

In the UK working late is a sign of dedication to the job, while in Germany it is viewed as a sign of poor time management.

Even something as simple as presenting a business card can be fraught with social pitfalls. In Japan, China and Korea there are different rules on how and when to present cards, what the card should contain, and even when you should examine cards you have been given. Get it wrong and you may have torpedoed your business meeting before you even start talking!

Social and religious customs

Dress code, eating habits and the rules of social etiquette vary widely across the globe. Behaviour you wouldn't think to question may be viewed as deeply offensive in some cultures, and you may also find the reverse is true.

It is all too easy to make mistakes that can have awkward or even serious consequences for effective communication. Do acquaint yourself with the customs and moral codes of the region you are going to, or dealing with. If you haven't travelled much, a little preparation can also help reduce the impact of culture shock.

English as a second language

English is spoken and understood, at least to some degree, in most parts of the world. Communicating with colleagues from other countries is therefore usually straightforward. But remember that speakers for whom English is not a mother tongue will appreciate help:

- Try not to speak too fast. While stressing every single syllable may be unnecessary and even patronizing, a clear and well-paced delivery will help ensure your meaning is understood.
- Don't use slang or colloquialisms unless you are sure the other party understands them.
- Paradoxically, technical terms and jargon may be more clearly understood by specialists than by non-engineering compatriots from your own culture. But once again, make sure people know what you are talking about.

Top tips – working with global teams

✔ Make allowances for local conditions when communicating with staff

✔ Find out about different cultures before you travel

✔ Be sensitive to differences in the local work ethic

✔ Do not assume that your culture has all the answers

✔ Be prepared to learn from your counterparts in other cultures

✔ Delay your reactions until you are sure that they are appropriate

13. Communicating science

Many institutions devote extensive time and resources to encouraging better public understanding of how science works. Science weeks, open days and specialist science channels and media programmes all help make research easier to grasp for the non-technically minded.

Explaining science

If you are asked to take part in a science education initiative:

- Target your audience. For example, younger school students will require a different approach to a group of scientifically literate adults.
- Keep things simple and do not assume prior knowledge of your subject. Focus on the needs of the least receptive member of the group — that way you will reach the widest possible number of people.
- Use visual examples, displays and demonstrations to make key points. Smells, bangs and audio-visual displays are more absorbing than a boring lecture, no matter how interesting the topic.
- Highlight the relevance of your subject to people's everyday lives and experience; not everyone is fascinated by science for its own sake.
- Tell people what the practical applications are. Only then attempt to interest them in the background science.
- Avoid using unnecessary technical terms. Impenetrable jargon will be an immediate turn-off for most people.
- Include the human angle. The story behind scientific discoveries can help people relate to the science itself.
- Do not assume the audience shares your enthusiasm. It is up to you to get them interested.

It is worth noting that not all branches of science and engineering have the same intrinsic public appeal. It all depends on how relevant the subject is to people's everyday lives. For example, medical breakthroughs are virtually guaranteed public and media attention. Conversely, it would be very difficult to interest non-scientists in, say, a development in pure mathematics that has no obvious practical applications.

Communicating uncertainty

Despite the best efforts of the educators, public confidence in some areas of science has declined to crisis levels in recent years. This lack of trust occurs when people think their health and safety is at risk and science cannot give a clear-cut answer, or scientists disagree.

People assume mechanics can solve problems with the car, accountants will sort out their tax affairs and doctors exist to make them better. In the same way, there is an expectation that the scientist's role is to make sense of natural laws and apply them to the betterment of mankind.

Of course, as a scientist or engineer you know there are times when science does not know the answers. Uncertainty is part and parcel of the scientific process. But non-scientists won't accept that.

Overcoming this problem is not as simple as educating the public about the way science works. Instead you need to take public concerns seriously and address them, even if they may be scientifically incorrect:

- Find out why people are worried.
- Say what you know and what you don't know.
- Deal with the issue; if you dismiss or ignore public concerns you will be seen as arrogant.
- Be seen to be doing something to investigate the problem and to find answers.

- Stress safety measures and redundancy in critical systems. Non-scientists may not know much about engineering but they are familiar with the workings of 'Sod's Law' (if something can go wrong, it will).
- Get it right! Operational competence is the best and easiest solution to public mistrust.

Risk perception and communication

Successful risk communication is not as easy as it might appear. The reason is that scientists and non-scientists interpret risk very differently.

To a risk analyst, risk is a function of two quantifiable factors: the potential magnitude of an event, and the statistical chances of the event occurring.

Non-scientists, on the other hand, assess risk by instinctively applying a number of different criteria to a potential hazard. Among the questions people ask themselves are:

- Did I agree to this?
- Do I have any say in the outcome?
- Have I been told everything about what's going on?
- Does the 'system' care about me?
- Am I going to be the fall guy if this goes wrong?
- If it does go wrong, can the consequences be controlled?

Working together

If the public is over-reacting to what is technically an insignificant risk, addressing some of these concerns can help to diffuse public anger and criticism. Don't ignore public concerns because they are 'unscientific'. Tackle them head on. For instance:

- Is it new? Explain the background.
- Does it seem scary? Educate and explain what's going on in everyday terms.

- Did I agree to this? Involve local communities in the debate.
- Do you know what you are doing? Explain the facts, openly and transparently.
- Does my opinion count? Show you care, and listen to relevant local knowledge.
- Can I trust you? Be open and don't hide anything.
- Are you sure you can reverse the situation if it gets out of control? Well, are you?

In the end, you may find openness, empathy and fairness more appropriate tools for normalizing public assessment of risk than any amount of scientific debate and education.

Top tips – communicating science

✔ Translate science into plain English, using everyday examples and analogies

✔ Explain why and how the science is relevant to people's lives

✔ Where possible, use visual aids and displays

✔ Be open and explain what you know and what you don't

✔ Treat public concerns over risk seriously

✔ Show you care

14. Talking to the media

Scientists who are good communicators may find journalists contact them for expert comment on developments in their field.

If you are under attack or involved in a controversy of some sort then you may be required to answer to public concern or criticism. In this case you might want to use the media to justify your position or defend your organization.

Newspapers, magazines, radio, television and the web-based news channels offer a powerful and influential means of communicating your organization's message to a broader audience. You can use the media to help you:

- Promote a product or project.
- Educate the public.
- Change the way people behave or think about an issue or subject.
- Influence your peers.

Most organizations have a press office to regulate contact with the media. Always refer to them or your manager before speaking to journalists in an official capacity.

The press release

If you are seeking publicity, the primary means of attracting media interest is the press release. This is simply a single sheet of paper containing a concise synopsis of whatever it is you want to say.

News desks receive dozens of press releases every day. To ensure yours gets used, keep things simple and focus on practical applications or implications. Most journalists do not have a scientific or engineering background and they will not have the time to grapple with technicalities.

To help you whittle your message down to its bare essentials, you may find it helpful to write just one sentence against each of the following headings:

- What?
- Why?
- How?
- Who?
- When?
- Where?

To complete the press release, add a short headline. This should ideally be phrased in such a way that it encourages the journalist to read further.

Make sure you include contact details of someone journalists can approach for more information if they need it.

If your press release contains all the necessary facts, journalists may use the story without referring back to you. But if a story is interesting they may ask you for an interview.

Handling media interviews

People worry about interviews, fearing the journalist may try and twist what they say, trap them or make them look foolish. In fact, a media interview usually works to your advantage. It is a chance for you to explain the story in your own way and in its proper context.

Preparing for an interview

Before the interview:

- Decide what you want to say.
- Anticipate obvious questions and think through possible replies.
- Make sure you have key facts and examples to hand.

Make sure the reporter understands the background to the story. If you can, try and have a chat before the interview starts.

What to expect in the interview

Newspaper reporters will interview you in person or over the telephone. Radio or television journalists will want to chat to you either in a studio, over the telephone, or on location (on-site). You may at some stage come across the so-called 'down the line' interview, which involves being interviewed remotely by a reporter in a distant studio.

Once the interview starts:

- Reply to questions, but make sure you say what you want to say as well.
- Don't assume prior knowledge on the part of the journalist or audience.
- Keep your arguments simple and easy to understand.
- Make the most of your voice, talk naturally and don't rush.
- Maintain strong eye contact with the reporter; it will give you more authority.
- Avoid technical jargon; the general public won't understand it. Use everyday English.

Avoid technical jargon in media interviews

Awkward interviews

If things have gone wrong you can expect the media to give you a hard time. Being grilled by an assertive journalist is an uncomfortable experience, especially if you know you or your organization is at fault. If you find yourself in this situation:

- Never say 'No comment!' It will make your organization appear uncaring.
- Prepare well and think through constructive answers to predictable questions.
- If there are good reasons for an action or policy, explain them.
- If the controversy or criticism arises from a misunderstanding, then use the media to educate the public and put the matter into its proper context.
- Be sure to stress steps you have taken to address or remedy the issue.
- Be prepared to be honest; you may win public support.

The soundbite culture

There is rarely time or enough space on the page to go into great depth on an issue. Journalists therefore like to boil complex issues down to the barest minimum of facts and possible interpretations.

This can be very frustrating for scientists and engineers who know that the natural world is complex and that conclusions drawn from data are rarely absolute. But knowing how to deliver an effective soundbite — a brief quote or summary — will save you time and give you more control over what is published or broadcast.

- Keep it short, 30 seconds at most (about 90 words).
- Focus on practical results or implications.

- Distil your message down to the barest essentials; three or four short sentences will do.
- Try not to be too pedantic — if a conclusion is a reasonable one to draw, then do so.
- If you want publicity and the opportunity of a longer interview is offered, take it. Insisting on giving the journalist a snappy quote would in this case be counter-productive.

Media training

Handling media interviews is a practical skill. Before you tackle the real thing, you may wish to consider taking a media training course. This will give you the chance to try your hand at realistic interview exercises in a safe and confidential learning environment.

Media training courses are generally run by ex-journalists or public relations companies. Be aware that it is an unregulated industry, and some tutors are better qualified than others. Before booking a course, check the trainer's professional experience and ascertain whether they can offer you the level of training you need.

Top tips – talking to the media

✔ Always clear media contact with your manager

✔ Translate technical jargon into plain English

✔ Prepare what you want to say before the interview

✔ Anticipate obvious questions

✔ Remember to put *your* points across

✔ Make the most of how you look and sound

15. Security and the law

Information is power and communication is a powerful tool. Both can be abused, so confidentiality and security of information are important issues. Communication can also have legal implications. Here are some frequently asked questions.

Communication and the law

(NB This section is intended as a general guide in the light of relevant English and US law current at the time of writing only. Always take professional legal advice.)

Q *What can I do if someone says something hurtful or damaging about me?*

A People can say what they like about you and there's nothing you can do about it so long as it is true. If it's not true and you can prove what they are saying has damaged your reputation — which can be difficult — you could bring a legal action for slander. By the same token, others could sue you if you are the source of false information.

Q *What could happen if I write something nasty about someone else?*

A If what you say isn't true and the other party can prove what you wrote has affected their reputation or that of their organization, you could be sued for libel. Libel includes anything that is disseminated widely and held on permanent record. Damaging false statements in the press and media broadcasts fall into this category.

Q *What about emails and other electronic communications?*

A There is a trend for emails to be treated in law like any other written communication, although this is a grey area due to the newness of the medium. False and damaging content that refers to others may therefore be actionable. Employers may be held responsible for the content of communications sent by their staff. You can protect yourself to some degree by adding a standard disclaimer to every email you send, but the safest solution is to be careful what you write.

Q *Can I use material off the web?*

A The ease with which words and pictures can be copied from the web means it is easy to fall foul of copyright law. There is a distinction between personal and business use. Always ask permission of the writer / site administrator before using someone else's material, particularly if this use will bring you commercial gain.

Q *I have been receiving unwelcome emails of a sexual nature. What can I do?*

A The sending of inappropriate email may be regarded in law as harassment, or even a criminal offence. Refer to your line manager or organization's personnel support team.

Q *How important is confidentiality?*

A You should regard any unpublished information relating to your employer's affairs as confidential and take reasonable care to ensure it stays that way.

Q *Do I have to worry if I hold information about others on my computer?*

A Under English law those processing personal data by computer must comply with eight enforceable principles of good practice. Data must be:

- Fairly and lawfully processed.
- Processed for limited purposes.
- Adequate, relevant and not excessive.
- Accurate.

- Not kept longer than necessary.
- Processed in accordance with the data subject's rights.
- Secure.
- Not transferred to countries without adequate protection.

Personal data covers both facts and opinions about the individual. Businesses are required to comply with the Data Protection Act 1998.

Q *Can my employer monitor the emails I send and the websites I look at?*

A This area of law is still in development. Employers may claim the right to monitor email traffic originating from their computer equipment and check what websites have been accessed from their computers, although in law they should be able to give reasonable reason for doing so. Be aware your cheeky email or surreptitious employment application could be intercepted, with disastrous results for your job! It is normal practice for employers to advise employees that they intend to monitor email traffic and web surfing.

Q *What if I say something wrong in a media interview?*

A Both you and the journalist could be sued for libel if you publish or broadcast false information that damages the reputation of another party. Journalists should be trained to watch for legal pitfalls like this, but they can't prevent you saying the wrong thing in a live television or radio interview.

Also, saying too much about a situation that is, or could in the future be, the subject of legal proceedings could prejudice those proceedings, and even lead to you being charged with contempt of court. Be careful and think through how much you want to say before you open your mouth. Don't give a media interview in a business situation without discussing it with your manager first.

Security

Q *How do I stop people within my organization gaining access to my files?*

A Start with common sense precautions like enabling password-only access on your PC and keeping sensitive files in a locked drawer or cupboard. Big organizations run multi-tiered computer access protocols that prevent workers seeing restricted or sensitive material unless they are cleared to do so.

Q *What about outsiders accessing my data?*

A There is a possibility 'hackers' may be able to get into a computer via an internet connection. Big organizations put elaborate security procedures called 'firewalls' in place to prevent this happening. If you're working on a stand-alone computer you should seriously consider installing a hard- or software firewall to prevent access.

Q *How about computer viruses?*

A Some computer viruses can access your personal data files and then send the contents to everyone in your address book. All big organizations protect their computer systems with anti-virus software. Small companies should regard the installation and regular updating of anti-virus software as a priority.

Q *What if I suspect someone is recording my telephone conversations?*

A Recording telephone calls made directly to you is common practice in the media, either for broadcast or to provide journalists with a record of what was said. Responsible journalists will tell you they are doing this (in fact it is an offence against the Telecommunications Act to make a recording without advising you first), but to play safe never say anything you might regret if it was published. If you suspect someone is secretly recording your phone traffic then inform your organization's security department or the police.

Q *I've heard that even my rubbish isn't safe...*

A It is possible for interested parties to snoop through an organization's rubbish for sensitive or useful information. A large organization will have a shredding policy in place. If you work for a small company, consider buying a personal shredder.

Q *How can I send secure emails?*

A You have two options. To ensure no-one can tamper with the contents, and to reassure recipients that the message came from you, you can 'sign' your message with a digital signature. Alternatively, you can encrypt your message. For more on email security consult your computer software help files or your IT administrator.

Q *What if I realize too late I've sent an email to the wrong person?*

A Always double check addressees before you send emails. If you sent the email to someone on the same network it may be possible to recall or delete the message if it hasn't been opened. It is a good idea to attach a standard postscript to all emails pointing out that the communication is private and asking the recipient to let you know if they have received your mail in error and then delete the message from their files.

Top tips – security and the law

✔ Treat your employer's affairs as confidential

✔ Be careful what you write about people

✔ Ensure emails go to the right address

✔ Only use your office computer for work

Further reading

Books

If you want to find out more about the subjects covered in this book, the following publications may be useful. This list is not exhaustive. For the latest titles consult a good bookseller's website, for example Amazon.com (www.amazon.com) or Waterstones (www.waterstones.co.uk).

The basics

101 Ways to Improve Your Communication Skills	Jo Condrill	GoalMinds, 1999
The MIT Guide to Science and Engineering Communication: 2nd edition	James Paradis & Muriel Zimmerman	MIT Press, 2002

The written word

Essential English	Harold Evans	Pimlico, 2000
How to Write Effective Reports	John Sussams	Gower, 1998

Telephone skills

Telephone Techniques	Lin Walker	AMACOM, 1999
Telephone Skills	Patrick Forsyth	CIPD, 2000

E-communications

Perfect E-mail	Steve Morris	Random House, 2000
Writing Effective E-mail	Nancy & Tom Flynn	Kogan Page, 2000

Body language

Body Language	Allan Pease	Sheldon Press, 1997
Teach Yourself Body Language	Gordon Wainwright	Hodder & Stoughton Educational, 1999

Job applications / job interviews

Write a Great CV	Paul McGee	How To Books, 2001
Effective Interviews	Jenny Rogers	AMACOM, 1999

Meetings

Essential Managers: Managing Meetings	Tim Hindle	Dorling Kindersley, 1998
How to Manage Meetings	Alan Barket	Kogan Page, 2002

Handling conflict

Handling Conflict and Negotiation	Manchester Open Learning	Kogan Page, 1993
Quick Skills: Handling Conflict: Learner Guide	Douglas Gordon	South-Western, 2000

Public speaking / Audio-visual aids

How to Develop Self-confidence and Influence People by Public Speaking	Dale Carnegie	Vermilion, 1990

What's Your Point? The 3-Step Method for Making Effective Presentations	Bob Boylan	Adams Media, 2001

Working in global teams

The Survivor's Guide to Business Travel	Roger Collis	Kogan Page, 2001
Rules of the Game: Global Business Protocol	Nan Leaptrott	Thompson Executive Press, 1995

Communicating science

Science in Public: Communication, Culture, and Credibility	Jane Gregory & Steve Miller	Perseus, 2000
Responding to Community Outrage: Strategies for Effective Risk Communication	Peter Sandman	AIHA Press, 1993

Talking to the media

Handling Publicity the Right Way	John Venables	Elliot Right Way, 1997
How to Handle Media Interviews	Andrew Boyd	Management Books, 2000

Websites

Websites disappear or change addresses over time. Specific sites are therefore not listed here. Use a good search engine such as Google (www.google.com) to find up-to-date and relevant information on the topic you are looking for.

It is worth checking out the STEMPRA site (www.stempra.org.uk). The Science, Technology, Engineering and Medicine Public Relations Association offers useful advice on communicating science, as well as links to other web addresses that may be of interest.

Communication and management skills online

You've read the book, now take it one stage further — IChemE can offer you 20 courses ranging from 'Giving Presentations' to 'Facilitation Skills'.

What's more, these courses are available as e-learning so you can study at a time and pace to suit you.

Priced at £49 to non-members (member discounts apply), the following courses are available:

Budgeting Basics
Budgeting is a complex process; this course gives a good grounding in the basic knowledge needed.

Managing Your Budget
Ensure that the plans you have for your budget can be met.

Preparing Your Budget
Ensure that you can write a realistic budget.

Appraisal Interviewing
Learn to improve the effectiveness of appraisal interviews, a commonly used process in many companies.

Coaching Skills
Learn the skills needed to improve your staff's aptitude and become a better manager.

Mentoring Skills
Develop the skills needed to make a successful mentor, or gain more from the process as a mentee.

Training for Non-Trainers
Learn the best ways to progress your staff.

Facilitation Skills
Learn to develop a more effective working team, whatever the project.

Managing Meetings
Learn to improve the effectiveness of personnel leading and participating in meetings.

Negotiation Skills
Develop important life and business skills in reaching mutually beneficial agreements.

Giving Presentations
Learn to present with confidence.

Preparing Presentations
Learn to prepare more structured and successful presentations.

Managing Yourself
Learn to work more effectively and feel more fulfilled at work.

Plan Your Own Development
Learn to improve skills in developing and planning continuous professional development.

The Assertive Manager
Learn to impose authority without resorting to aggressive behaviour.

The Effective Leader
Learn to improve leadership skills for better management.

Team Building
Learn to build and maintain more effective teams.

Problem Solving
Learn to deal with problems effectively by producing better solutions.

Project Management
Learn practical techniques for improving the management of projects and meeting deadlines.

Report Writing
Learn to plan and produce effective reports.

Visit www.icheme.org/learning for more details — here you will also find IChemE's full range of technical e-learning courses covering important safety issues.

To discuss further please call IChemE on +44 (0) 1788 578214.

All prices correct at time of going to press but are subject to change without prior notification.

Management courses provided courtesy of KnowledgePool.

E-learning gives you the freedom and flexibility to develop a new skill and enhance your work practices. It gives you the opportunity to take control of your career advancement by offering affordable training in communication and management skills as well as a range of technical and safety needs.

In addition to the online services offered, IChemE also provides a full range of training courses for the chemical and process industries, including 'Communicating Risk' and 'Facing the Media'. For your free copy of the training brochure, contact the Events Department on +44 (0) 1788 578214 or email: courses@icheme.org.uk, or visit www.icheme.org/learning. If you are a training provider and wish to be included within our programme, please contact us with details of your courses.

RECEIVED
4 2003